David Ramsay, William Price Young

A Sketch of the Soil, Climate, Weather and Diseases of

South-Carolina

David Ramsay, William Price Young

A Sketch of the Soil, Climate, Weather and Diseases of South-Carolina

ISBN/EAN: 9783337345754

Printed in Europe, USA, Canada, Australia, Japan

Cover: Foto ©berggeist007 / pixelio.de

More available books at **www.hansebooks.com**

A

SKETCH

OF THE

SOIL, CLIMATE, WEATHER,

AND

DISEASES

OF

SOUTH-CAROLINA,

READ BEFORE THE MEDICAL SOCIETY OF THAT STATE,

BY

DAVID RAMSAY, M. D.

VICE-PRESIDENT OF THE SOCIETY.

CHARLESTON:
PRINTED BY W. P. YOUNG,
FRANKLIN'S HEAD, NO. 43, BROAD-STREET.

MDCCXCVI.

SKETCH

OF THE

SOIL, CLIMATE, WEATHER, AND DISEASES

OF

SOUTH-CAROLINA.

SOUTH-CAROLINA nearly refembles a tri-
angle—It is bounded on the eaft by the Atlantic ocean, and extends thereon about two
hundred miles; on the fouth, and partly on
the weft by the river Savannah; and on the
north, and partly on the weft by North-Caro-
lina. Thefe two laft mentioned boundary
lines approximate to each other, about three
hundred miles from the fea-coaft, and in the
vicinity of the Alleghany mountains.

The ftate of South-Carolina lies between
the 32d and 35th degrees of north latitude.
Its chief city, Charlefton, is in north latitude
32° 45, and in weft longitude from London,
79°, and from Philadelphia, 5°, and ftands on
a point of land between the junction of Afhley and Cooper rivers, and about ten miles
from the ocean.

In treating of South-Carolina, the philo-
fopher, as well as the politician, muft confi-
der it as divided into upper and lower country. Nature has marked this diftinction in
many particulars. Along the fea-coaft, and
for one hundred miles weftward, the country
is generally low and flat; from thence, to its
weftern extremity, it is diverfified with hills,
rifing higher and higher, till they terminate in
the Alleghany mountains, which are the part
age ground of the eaftern and weftern waters.
In the vallies, between thefe hills, a black and
deep loam is found. This has been formed by
abrafion from the hills, and from rotten trees
and other vegetables, which have been col-
lecting for centuries.

The rivers of the upper country originate
in the mountains, and are an affemblage of
ftreams. After thefe have paffed into the low
country, they move flowly, and in a ferpen-
tine courfe, till they empty into the ocean.
The rivers of the low country are, properly,
arms of the fea, extending but a few miles
till they head in fwamps and marfhes.

Carolina, lying on the eaft fide of the part-
age ground, between the eaftern and weftern
waters, is confiderably lower than the corref-
ponding parts of the United-States, which
are on its weft fide. Hence it follows, that
when the fnows melt, or heavy rains fall on
the mountains, much more of the water,
                                        proceeding

proceeding from thefe fources, is determined to the Atlantic ocean than to the river Miſſiſſipi. In confequence of which, we are often too wet, while our weſtern neighbours are too dry.

The ſide of South-Carolina, which borders on the fea, is interfected by thirteen rivers, viz. The Waccamaw, Black-river, Santee, Wandow, Cooper, Aſhley, Stono, Ediſto, Aſheppoo, Combahee, Coofaw, Broad, and May rivers. Some of thefe have two mouths, others have feveral heads, or branches. The river Santee, in particular, is formed by a junction of the waters of the Enoree, Tyger, Pacolet, and Catawba rivers, which originate in the mountains. All of the firſt mentioned thirteen rivers have a margin of fwamp always on one ſide, but often on both, extending from half a mile to three miles.

Thefe fwamps, in their natural ſtate, abound with ufeful timber of various kinds, and, when cleared, they reward their cultivators' with plentiful crops, efpecially in feafons that are exempt from frefhes. In the intervals between thefe rivers, there are often inland fwamps, frefh-water lakes, and great quantities of low level land, which, after heavy rains, continue for a long time overflowed. The remainder is a dry, and, for the moſt part, a fandy foil.

The foil of South-Carolina is naturally, and, for the purpofes of taxation, politically divided into the following claffes. 1, Tide-fwamp. 2, Inland fwamp. 3, High river fwamp, or low grounds, commonly called fecond low grounds. 4, Salt marfh. 5, Oak and hickory high land. 6, Pine barren. The tide and inland fwamps are peculiarly adapted to the culture of rice and hemp. The high river fwamps to hemp, corn, and indigo. The falt marfh has hitherto been, for the moft part, neglected; but there is reafon to believe, that it would amply repay the expence and labour of preparing it for cultivation. The oak and hickory high land is well calculated for corn and provifions, and alfo for indigo and cotton. The pine barren is the leaft productive fpecies of our foil, but it is the moft healthy. Daily experience proves that, under certain circumftances, it may be cultivated to advantage for provifions, indigo, and cotton. A proportion of it is an indifpenfably neceffary appendage to a fwamp plantation. It is remarkable that ground of this laft defcription, though comparatively barren, affords nourifhment to pine trees, which maintain their verdure through winter, and adminifter more to the neceffities and comforts of mankind than any other trees whatfoever. This may perhaps, in part, be accounted for by the well-known obfervation, that much of the pine land of this ftate is only fupcrficially fandy, for by digging into it a few inches or feet, the

foil,

foil, in many places, changes from fand to clay. ..

In digging into the fwamps, on the margin of the rivers, the operator frequently meets with the trunks of large trees, which appear to have been buried for ages, and is always arrefted in his progrefs by the fpringing of water. As deep as thefe fwamps have been penetrated, they confift of a rich blue clay, or a black foft mould, of inexhauftable fertility.

From this defcription of the low country, it is apparent, that there muft be a predominance of moifture; and from the co-operation of heat, there is a ftrong tendency to putrefaction. From the fame caufes, and the prefence of acid gafes, floating in the common atmofphere, metals are very fubject to ruft. This is particularly the cafe with iron, which, when expofed to the air, lofes, in a fhort time, all its brightnefs, and much of its folidity.

The climate of South-Carolina is in a medium between that of tropical countries, and of cold temperate latitudes. It refembles the former in the degree and duration of its fummer heat, and the latter in its variablenefs. In tropical countries, the warmeft and cooleft days, do not, in the courfe of a twelve month, vary more from each other than fixteen degrees of Fahrenheit's thermometer: there is, confequently, consequently, but little diſtinction between their ſummer and winter: but a variation of 83 degrees between the heat and cold of different days in the ſame year, and of 46 degrees in the different hours of the ſame day in South-Carolina, is to be found in its hiſtorical records.

In our cooleſt ſummers, the mercury in the thermometer* has reached 89, and in the five laſt years in which obſervations have been made by this ſociety, it has never riſen above 93, nor fell below 28. In the year 1785 it ſtood for a few hours at 96, which was its greateſt height ſince the year 1752, when it roſe to 101. In the year 1794 it was never lower than 34, during the time of obſervation, which began at eight in the forenoon, and ended at ten in the evening. The difference between our cooleſt and warmeſt ſummers, therefore, ranges between 89 and 96; and the difference between our mildeſt and ſevereſt winters, ranges between 34 and 28. Our greateſt heat is ſometimes leſs, and never much more, than what takes place in the ſame ſeaſon in Baltimore, Philadelphia, and New-York;

* Fahrenheit's thermometer is what is every where meant in this publication, and the obſervations on it, therein referred to, were reported to the medical ſociety, as taken by Dr. Robert Wilſon, at his houſe, the weſt end of Broad-ſtreet, at the hours of eight in the morning, between two and three in the afternoon, and at ten in the evening. The inſtrument was ſuſpended in an open paſſage, about ten feet from the earth.

York; but their warm weather does not, on an average, continue above six weeks, while ours lasts from three to four months. Our nights are also warmer than theirs. The days in Charleston are moderated by two causes, which do not exist, in an equal degree, to the northward of it. Our situation open and near the sea, almost surrounded by water, and not far distant from the torrid zone, gives us a small proportion of the trade winds, which blowing from the south-east are pleasantly cool. These generally set in about 10 A. M. and continue for the remainder of the day. A second reason may be assigned from the almost daily showers of rain that fall in the hottest of our summer months.

Since we began our meteorological journal (January, 1791) the mercury in the thermometer has never been under 28, though in the year 1752 it was down to 18. Mr. Hemitt, in his historical account of South-Carolina, asserts, that he had seen the mercury in Fahrenheit's thermometer, down to 16, and that others had observed it as low as 10. On the whole, for five years past, our greatest heat has been eight degrees, and our greatest cold ten degrees less than they were about the middle of this century, as observed by Dr. Chalmers. A similar observation, though not to the same extent, will result from comparing the greatest heat and cold of the five last years, 1791, 1792, 1793, 1794, and 1795,

1795, as recorded by the medical society, with the years 1750, 1751, and 1752, the three firft years recorded by Dr. Chalmers. The greateftheat in 1791, was 90, in 1792, 93, in 1793, 89, in 1794, 91, in 1795, 92; but the greateft heat in 1750, was 96, in 1751, 94, in 1752, 101. The greateft cold in 1791, was 28, in 1792, 30, in 1793, 30, in 1794, 34, in 1795, 29; but in the year 1750, it was 25, in 1751, it was 23, and in 1753, it was 18. Whether this change is accidental, or the confequence of an improvement in our climate, time and future obfervations muft determine. The advantages refulting to the temperature of the air, and to the healthinefs, as well as to the appearance of any country, from the art of man, inhabiting and cultivating it, are inconceivably great. We may, therefore, indulge the hope, that ours is progreffively meliorating from permanent and encreafing caufes.

The quantity of low and moift ground in Carolina, is daily diminifhing. Cultivation naturally tends to exficcation. Wherever the tide flows it brings fomething with it, which being left, helps to fill up cavities. Indeed the furface of the earth naturally, and univerfally, approximates to a level. The rains wafh from the high grounds, and add what is carried away to the low. The bones of an enormoufly large animal have been lately dug up in Biggin-fwamp, by the la-
bourers

bourers at the Santee Canal, eight feet under ground. The trunks of trees have been frequently found at an equal or greater depth. It is poffible that thefe may have been buried below the furface of the ground, as deep as they were lying, but it is much more probable that they originally funk in the earth, one, two, or three feet, by their own weight, and were afterwards covered by fucceffive alluvions in the lapfe of time, to the depth at which they were found.

In proportion as our country has been cleared and cultivated, its rich low grounds, from various caufes, have become higher and drier. Much fand and dry clay has been blown on them by high winds. The cutting down of trees has deftroyed their perfpiration. Many hundred gallons of water are daily iffuing from every acre of ground that is fully timbered. The exhalation from the bare furface of the earth expofed to the fun, is much greater than it would be, if the fame ground was covered with trees. It is a well known fact, that many old rice fields are now much lefs productive, than they were thirty years ago. It is probable, that the day is not far diftant, when much of the fwamp of this ftate, will be converted into dry arable land, more fit for corn than rice. Though the moifture of the foil has in general decreafed, with our increafing cultivation, yet frefhes in fuch of our rivers, as originate in the mountains, have, for fome

c   years

years paft, been higher, and more frequent than ufual.

Thefe are ferious evils, threatening the def-truction of fome of our moft valuable lands. To inveftigate the caufes thereof, is an object well worthy the attention of every friend to Carolina. One reafon affigned for the late in-creafe of frefhes is, that the clearing of the upper country opens many fprings, and gives circulation to much of what would, in a ftate of nature, be ftagnant water. By means of drains, made with a view of rendering the ground plantable, the water, which would other-wife remain quiefcent, till it was either abfor-bed, or evaporated, is conducted to the near-eft ftream, all of which, fooner or later, emp-ty into the rivers. It is within the recollec-tion of the old inhabitants of our upper coun-try, that the rivers thereof were, in the days of their youth, much more fhallow than they are at prefent. If the obfervation already made, " That the tide, wherever it flows, brings fomething with it, which being left behind, helps to fill up cavities," is well found-ed, may we not fuppofe, that the floods, rufh-ing down the rivers from the mountains, meet with obftructions, yearly increafing, which retard their courfe to the ocean? If this is one caufe, among others, of the increafe of frefhes, the remedy would be to expedite the paffage of the water from the rivers to the fea, by multiplying and enlarging their vents,

and

and fhortening their courfe. Whether this is
practicable to an extent that would fave all
the land adjacent to the rivers, is very doubt-
ful; but it certainly might be effected fo as to
fave many plantations, provided the owners
would fyftematically co-operate in the execu-
tion of a judicious plan, for the more fpeedy
difcharge of the fuperfluous water.

The common tides in Afhley and Cooper
rivers rife in Charlefton from fix to eight feet;
the fpring tides from eight to ten. A common tide, with an eaftwardly wind, is higher
than a fpring tide, with a weftwardly wind.
The tides in general afcend our rivers about
thirty five miles from the ocean, in a direct
line. The higheft ground in Charlefton, is
between nine and ten feet above the higheft
fpring tides. This is to be found in George-
ftreet, between Meeting and King ftreets. The
next higheft ground is in Harlefton, in Went-
worth-ftreet. The next in the weft end of
Broad-ftreet, near the theatre. The next in
Meeting-ftreet, nearly oppofite the new mar-
ket.

Earthquakes are fo rare, and fo flight, as not
to have been noticed in our hiftorical records.
A momentary one, that did no damage, is re-
collected by fome of our old citizens, as having
taken place about the middle of the prefent
century. But whirlwinds are more common.
Thefe, for the moft part, are confined to narrow

limits, and run in an oblique direction, level-
ling the loftieft trees that ftand in their way.

   There are fome circumftances which make
it probable, that the whole of the low country in Carolina, was once covered by the ocean.
In the deepeft defcent into the ground, nei-
ther ftones nor rocks obftruct our progrefs,
but every where fand or beds of fhells: inter-
mixed with thefe, at fome confiderable depth
from the furface, petrified fifh are fometimes
dug up. Oyfter fhells are found in great quantities, at fuch a diftance from the prefent limits
of the fea-fhore, that it is highly improbable
they were ever carried there from the places
where they are now naturally produced. A
remarkable inftance of this occurs in a range
of oyfter-fhells extending from Nelfon's ferry,
on the Santee-river, fixty miles from the ocean,
in a fouth-weft direction, paffing through the
intermediate country, till it croffes the river
Savannah, in Burke-county, and continuing
on to the Oconee-river, in Georgia. The
fhells in this range are uncommonly large, and
are of a different kind from what are now
found near our fhores. They are in fuch
abundance, as to afford ample refources for
building and agriculture. At the diftance of
fix, eight, or ten feet from the furface, near
our fea-coaft, water univerfally fprings. A
fmall proportion of fea falt is found in all the
well water of this city, and it is probable
that the whole of it is obtained by filtration
from the ocean, or adjacent rivers.   Our

Our country partakes fo much of the nature of a Weft-India climate, as to be liable to hurricanes, but thefe have been lefs frequent than formerly. Within the firft fifty-two years of the prefent century, three took place, viz. in 1700, 1728, and 1752, but for the laft forty-three years nothing of the kind, worthy of notice, has occurred. Our elder citizens inform us, that thunder ftorms were, in the days of their youth, much more frequent and more injurious than they have been for the laft thirty years. This is remarkably the cafe in Charlefton, and is probably, in part, owing to the multiplication of electrical rods. Mr. Hewitt, who wrote about twenty-five years ago afferts, that he had known in Charlefton five houfes, two churches, and five fhips ftruck with lightning, during one thunder ftorm. Nothing comparable to this has occurred for many years paft. It is neverthelefs true, that during the fummer, there are few nights, in which lightning is not vifible in fome part of the horizon.

The tranfitions from heat to cold are great, and fometimes very fudden. Dr. Chalmers ftates, that on the 10th of December, 1751, the mercury in Fahrenheit's thermometer fell forty-fix degrees in fixteen hours, that is, from 70 to 24. The greateft variation that has taken place in a day, in the five years that have paffed fince the inftitution of this fociety, was on the 28th of October, 1793, when it fell to

37

37 from 74, at which it ftood on the 27th;
that is thirty-feven degrees in the courfe of
twenty-four hours.

The number of extreme warm days in
Charlefton is feldom above thirty in a year,
and it is rare for three of thefe to follow each
other. On the other hand, eight months out
of twelve are moderate and pleafant. The
number of piercing cold days in winter is
more, in proportion to our latitude, than of
thofe which are diftreffingly hot in fummer,
but of thefe more than three rarely come to-
gether. There are, on an average, in this
city, about twenty nights in a twelvemonth,
in which the clofenefs and fultrinefs of the
air forbid us, in a great meafure, the refrefh-
ment of found fleep, but this fevere weather
is, for the moft part, foon terminated by re-
frefhing and cooling fhowers. April, May, and
June are, in common, our healthieft months;
Auguft and September the moft fickly; April
and May the drieft; June, July, and Auguft
the wetteft; November the pleafanteft. In
fome years January, and in others February
is the coldeft month. It is remarkable, that
when orange trees have been deftroyed by froft,
it has always been in the month of February.
December is the beft month in the year for
ftrangers to arrive in this city: fuch fhould
calculate fo as not to make their firft appearance either in fummer, or the two firft months
of autumn. The hotteft day of the year is
                                          fometimes

fometimes as early as June, which was the
cafe in the year 1791; fometimes as late as
September as in the year 1793; but oftenest
in July or Auguft. The hotteft hour of the
day in Charlefton varies with the weather:
it is fometimes as early as ten in the forenoon,
but moft commonly between two and three
in the afternoon.

In the fpring when the fun begins to be pow-
erful, a langour and drowfinefs is generally
felt, refpiration is accelerated, and the pulfe
becomes quicker and fofter. Strangers are apt
to be alarmed at thefe feelings, and anticipate
an increafe of them, with the increafing heat
of the feafon, but they find themfelves agreeably
difappointed. The human frame fo readily ac-
commodates itfelf to its fituation, that the
heat of June and July is, to moft people, lefs
diftreffing than the comparatively milder weather of April and May. On the other hand,
though September is cooler than the preceeding months, it is more fickly, and the heat of
it more oppreffive. Perfpiration is diminifhed
and frequently interrupted; hence the fyftem,
debilitated by the fevere weather of July and
Auguft, feels more fenfibly, and more frequently, a fenfe of languor and laffitude. Befides the
coolnefs of the evenings in September, and
the heavy dews that then fall, multiply the
chances of getting cold. It is, on the whole,
the moft difagreeable month in the year.

Frofts

Frosts feldom extend into the ground more than two inches in the coldeft feafons. They generally commence about the middle of October, and terminate in the month of March. On their approach they bring with them a cure for the fevers then ufually prevalent. The inhabitants of Charlefton keep fires in their houfes from four to fix months in the year, but there are fome warm days in every month, in which fires are difagreeable. On the other hand, there are fome moift cool days in every month of the year, with the exception of July and Auguft, in which fires are not only healthy but pleafant. Ice is feldom half an inch thick, and rarely gives an opportunity for the wholefome exercife of fkating.

The annual medium temperature of the air in Charlefton, was $65\frac{2}{12}$ in 1791, 65 in 1792, $65\frac{2}{12}$ in 1793, 65 in 1794, $64\frac{5}{12}$ in 1795. The average medium for thefe five years, without fractions, is 65. The average medium of the ten years, viz. from 1750 to 1759, which were obferved and recorded by Dr. Chalmers, was 66. From thefe facts it appears probable, that the aggregate heat of different years, in the fame place, is nearly equal. A very warm fummer is preceeded or followed by a proportionably cold winter, fo as to bring different years nearly to the fame temperature of the air, on an average of the whole four feafons.

The greateft, leaft, and mean heat, for every month of the year, for the five laft years, will appear from the annexed table.

TABLE OF THE GREATEST, LEAST, AND MEAN DEGREES OF HEAT, IN CHARLESTON, FOR THE YEARS

Month.	1791	1792	1793	1794	1795
January	G. 65 L. 35 M. 50	G. 66 L. 30 M. 48	G. 67 L. 36 M. $51\frac{1}{2}$	G. 65 L. 35 M. 50	G. 60 L. 33 M. $46\frac{1}{2}$
Febru.	G. 69 L. 35 M. 52	G. 68 L. 30 M. 49	G. 74 L. 35 M. $54\frac{1}{2}$	G. 70 L. 34 M. 52	G. 63 L. 29 M. 46
March	G. 78 L. 42 M. 60	G. 74 L. 41 M. $57\frac{1}{2}$	G. 72 L. 34 M. 53	G. 76 L. 43 M. $59\frac{1}{2}$	G. 73 L. 33 M. 53
April	G. 82 L. 52 M. 67	G. 80 L. 52 M. 66	G. 83 L. 56 M. $69\frac{1}{2}$	G. 74 L. 50 M. $62\frac{1}{2}$	G. 78 L. 53 M. $65\frac{1}{2}$
May	G. 87 L. 61 M. 74	G. 84 L. 64 M. 74	G. 83 L. 62 M. $72\frac{1}{2}$	G. 86 L. 63 M. $74\frac{1}{2}$	G. 84 L. 70 M. 77
June	G. 87 M. 69 L. 78	G. 89 L. 63 M. $76\frac{1}{2}$	G. 86 L. 70 M. 78	G. 91 L. 65 M. 78	G. 86 L. 71 M. $78\frac{1}{2}$
July	G. 89 L. 66 M. $77\frac{1}{2}$	G. 93 L. 70 M. $81\frac{1}{2}$	G. 88 L. 76 M. 82	G. 85 L. 72 M. $78\frac{1}{2}$	G. 92 L. 74 M. 83
Auguſt	G. 90 L. 74 M. 82	G. 92 L. 69 M. $80\frac{1}{2}$	G. 87 L. 70 M. $78\frac{1}{2}$	G. 91 L. 75 M. 83	G. 88 L. 72 M. 80
Sept.	G. 87 L. 61 M. 74	G. 85 L. 60 M. $72\frac{1}{2}$	G. 89 L. 69 M. 79	G. 88 L. 66 M. 77	G. 83 L. 59 M. 71
Oct.	G. 83 L. 50 M. $66\frac{1}{2}$	G. 77 L. 46 M. $61\frac{1}{2}$	G. 82 L. 35 M. $58\frac{1}{2}$	G. 75 L. 47 M. 61	G. 79 L. 48 M. $63\frac{1}{2}$
Novem.	G. 72 L. 40 M. 56	G. 74 L. 45 M. $59\frac{1}{2}$	G. 76 L. 39 M. $57\frac{1}{2}$	G. 74 L. 37 M. $55\frac{1}{2}$	G. 75 L. 42 M. $58\frac{1}{2}$
Decem.	G. 63 L. 28 M. $45\frac{1}{2}$	G. 70 L. 34 M. 52	G. 66 L. 30 M. 48	G. 68 L. 37 M. $52\frac{1}{2}$	G. 71 L. 30 M. $50\frac{1}{2}$

To face Page 16.

The evils that every year take place, more or lefs, in Philadelphia, from drinking cold water, are unknown in this city. Our water lies fo near the furface of the earth, that the difference of its temperature from that of the common air, is not fo great as to create danger, unlefs in very particular circumftances. A folitary cafe occured in September, 1791, of a negro fellow, who, after taking a draught of cold water, when very warm, fuddenly fainted away, and, immediately after, became infane, and continued fo for feveral days, but he afterwards recovered.

Inftead of fudden deaths from cold water, we have to lament the fame event from the intemperate ufe of fpirituous liquors. The ftimulus of ardent fpirits, added to the ftimulus of exceffive heat, drives the blood forcibly on the brain, and produces fatal confequences. Thefe are oftener apoplexies than ftrokes of the fun. Four fots expired fuddenly, in one hot day laft fummer, in one fquare of this city.

The caft and noth-eaft winds in winter and fpring, are very injurious to invalids, efpecially to thofe who have weak lungs, or who are troubled with rheumatic complaints. In thefe feafons they bring with them that languor, for which they are remarkable in other countries; but in fummer, by moderating heat, they are rather wholfome than otherwife.

D 2 Weft

Weſt and north-weſt winds, which blow over large tracts of marſh, are, in the ſummer ſeaſon, unfriendly to health. The north and north-weſt winds are remarkable for their invigorating effects on the human frame. South winds are healthy in ſummer, but much leſs ſo in winter.

The general direction of the winds in this city, for four ſucceſſive years, may be learnt from the annexed table.

On December 31, 1790, at four o'clock, A. M. wind N. E. a ſevere ſnow ſtorm began in Charleſton, which continued for twelve hours, in conſequence of which, the ſtreets were covered with ſnow, from two to four inches deep, and the ſea iſlands, north-eaſtward, to the depth of ſix inches. Another took place on the 28th of February, 1792, wind N. W. which continued for ſeveral hours, and till it covered the ground five or ſix inches. Theſe were rare phænomena. Snow is more common, and continues longer in proportion as we recede from the ſea-ſhore. The further we proceed weſtward, till we reach the mountains, which divide the weſtern from the eaſtern waters, the weather is colder, and vegetation later. While the inhabitants of Charleſton can ſcarcely bear to be covered, in the hours of ſleep, with a ſheet, they who live in the town of Columbia, one hundred and twenty miles to the north-weſt

of

TABLE of the COURSE of the WINDS.

Month.	1791 Wind.	Days.	1792 Wind.	Days.	1793 Wind.	Days.	1794 Wind.	Days.
January	N. E. & E.	7	W. & N. W.	13	N. W. & W.	18	N. W.	9
	S. W. & W.	3	S. W.	2	S. E. & E.	4	N. E.	10
	N. W.	1	N. E.	15	S. W. & S.	8	W. & S. W.	12
					N. E. & E.	6		
February	S. E.	3	W. & N. W.	15	N. E. & N.	7	N. E.	6
	S. W. & W.	9	S. W.	4	N. W. & W.	12	N. W.	11
	N. E. & E.	12	N. E. & E.	11	S. E. & E.	8	S. W. & W.	10
	N. W.	4			S. W. & S.	7		
March	N. E. & E.	12	W. & S. W.	12	N. E. & E.	9	S. W. & W.	13
	S. W. S. & W.	14	N. W.	6	S. W. & W.	17	S. E. & S.	11
	N. W.	2	N. E. & E.	5	N. W. & N.	6	N. W. & N.	4
			S. E.	3	S. E.	7	N. E. & N.	11
April	S. E. & S.	5	S. E. & E.	4	N. E. & E.	14	N. E. & E.	14
	S. W. & W.	14	S. W. & W.	21	S. E. & S.	5	S. W. & W.	3
	N. E. & E.	14	N. W.	2	S. W. & W.	13	N. W.	5
					N. W. & N.	4	S. E. & S.	10
May	N. W & W.	4	W. & S. W.	13	S. E. & E.	14	N. E. & N.	10
	S. W. & W.	12	N. N. E & E.	12	S. W. & W.	9	S. E. & S.	15
	S. E. & E.	15	S. E. & E.	3	N. E.	10	S. W. & W.	12
	S. W.	1			N. W.	2	N. W.	7
June	S. E. & E.	10	S. W. & W.	13	S. W. & W.	17	N. W. & W.	14
	S. W. & W.	14	S. E. & E.	10	N.	1	S. E. & E.	7
	N. E.	2	N. E.	9	S. E. & S.	4	S. W. & S.	18
			N. W.	2			N. E.	3
July	S W. & W.	12	S. W. & W.	13	S. W. & W.	21	N. E. & E.	7
	N. W. & N.	3	N. E. & E.	16	S. E & S.	13	S. E. & S.	5
	N. E. & E.	14	S. E.	7	N. W.	2	S. W. & W.	20
	S. E.	3	N. W.	1	N. E. & E.	5	N. W. & N.	2
August	W. S. W. & S.	19	W. & S. W.	12	S. W.	14	N. E. & E.	16
	S. E. & E.	6	N. E. & E.	15	N. E. & E.	11	S. W. & W.	11
	N. E.	2	S. E.	5	N. W.	2	N. W. & W.	3
	N. W.	3	N. W.	4	S. E. & S.	11	S. E. & S.	10
September	S. W. & W.	12	N. E. & E.	23	S. W. & S.	9	N. E. & E.	14
	N. W.	6	S. E. & S.	1	S. E. & E.	10	S. E. & S.	8
	S. E. & S.	6	S. W. & W.	6	N. F.	19	S. W. & W.	15
	N. E.	8	N. W.	3	N. W.	5	N. W.	2
October	N. N. E. & E.	16	N. W. & N.	7	N. E. & E.	20	N. E. & N.	19
	N. W. & W.	14	S. W. & W.	4	N. W. & W.	7	N. W. & W.	5
	S. E.	3	N. E. & E.	18	S. E. & S.	10	S. E. & E.	10
					S. W.	4	S. W.	4
November	N. W.	12	N. W. & W.	18	S. W. & W.	10	S. W. & W.	9
	N.	3	S. W. & S.	7	N. E. & E.	10	N. E. & E.	13
	N. E.	7	N. E. & E.	10	S. E. & S.	6	S. E.	2
	S. & S. E.	2	S. E.	8	N. W. & N.	3	N. W.	8
December	W. N. W. & N.	22	N. E. & E.	10	N. W.	13	N. W. & N.	10
	N. E.	3	N. W. & N.	15	N. E. & E.	5	S. W. & W.	10
	S. W.	3	S. W. & W.	11	S. W. & W.	9	N. E. & E.	10
	S. E.	3	S. E.	1	N.	4	S. E. & S.	3

To face p. 18.

of it, are not incommoded with a blanket. The difference is greater as we advance to Ninety-fix, Pinckney, and Wafhington diftriéts.

The fum total of rain, on an average of ten years, viz. from 1750 to 1759, as obferved by Dr. Chalmers, was 41. 75 inches in the year. The quantity of rain that fell in each month of the year 1795, was as follows:

	INCHES.	10ths.
January,	8	5
February,	· 1	8
March,	4	6
· April,	2	4
May,	8	1
June,	8	1
July,	5	2
Auguft,	9	4
Sept. and Oétober,	8	9
November,	0	9
December,	5	0
	71	8 in the year.

In

In the four years preceeding 1795, before we began to meafure the quantity of rain, the number of days on which it fell in confiderable quantlties, without noticing flight tranfient fhowers, was as follows:

DAYS OF RAIN.

	1791	1792	1793	1794
January,	2	12	12	9
February,	8	7	9	5
March,	9	8	11	12
April,	6	2	9	7
May,	3	6	14	8
June,	15	9	8	13
July,	10	9	10	23
Auguft,	10	10	15	13
September,	10	6	8	9
October,	8	4	3	8
November,	9	5	9	10
December,	6	10	6	11
	96	88	114	118

When the waters are kept in motion by a fucceffion of fhowers, it is generally healthy; but fevers are ufually rife, when a feries of warm dry days follows great falls of rain. The ciftern water of this city, collected from rain, is a degree and a half warmer than the well water; and the temperature of the well water is 64½, which is twelve degrees warmer than that of Philadelphia.

Our

Our old people are ofteneft carried off in cold weather; the young, the intemperate, and the labouring part of the community, when it is hot.

It is to be regretted, that bilious remitting and intermitting fevers have increafed in the country, with the clearing thereof. The felling of trees, and opening of avenues to the rivers, have given more extenfive circulation to marfh miafmata. The increafe of mill-dams in the upper country has been injurious to the health of its inhabitants. In Charlefton a change has taken place much for the better. Bilious remitting autumnal fevers have, for fome years paft, evidently decreafed. The fmall-pox is now a trifling diforder, compared with what it was in 1760 and 1763. Pleurifies, which were formerly common and dangerous, are now comparatively rare, and fo eafily cured, as often to require no medical aid. The dry belly-ache has, in a great meafure, difappeared: perhaps this may be in part owing to the increafing difufe of punch. April and May ufed to be the terror of parents; but the difeafes, which thirty years ago occafioned great mortality among children in the fpring, have, for fome years paft, been lefs frequent and lefs mortal. It is now found, by happy experience, that they are often cured, or prevented, by country air. The three laft Aprils have paffed over without any notice being taken on our journals, of

the

the diarrhœa of infants, as having occurred in the practice-of the members of this society.

A species of fore throat, accompanied with symptoms of the croup, which formerly swept off numbers of children, has, for the four last years, rarely occurred in practice. More rational methods of treating wives and mothers, have been substituted in lieu of the enervating confinement, impofed in the days of our fathers. The good effects of which are vifible in the diminifhed number of women who die in childbirth, and in the increafing number of children who are now raifed to maturity.

Dr. Mofely, in his treatife on tropical difeafes, obferves as follows, " Hot climates are indeed very favorable to geftation and parturition. Difficult labours are not common, and children are generally born healthy and ftrong, and thrive more than they do in temperate climates, for a few years, and are not fubject to the rickets nor the fcrophula." As a proof of this general pofition, applied to our ftate, I obferve, that, in many inftances, from feven to ten, and in a few, from ten to fifteen children have been raifed to maturity in South-Carolina, from a fingle pair. There are now eight families in Broad-ftreet, between the ftate-houfe and the weftern extremity of that ftreet, in which fixty-nine children have been born, and of thefe fifty-fix are alive. In that

part

part of Meeting-ſtreet, which lies between Tradd-ſtreet and Aſhley-river, from ſix marriages, (which, with the exception of one, have taken place ſince the year 1782) forty-two children have been born, all of which, except three, are now alive, and the eldeſt of the whole is little more than fourteen. Within the ſame limits, ſeven other couple have fifty-two children living, the youngeſt of whom is twelve years old, and forty-ſeven are grown to maturity.

Greater inſtances of fœcundity frequently occur in our middle and upper country, chiefly among thoſe who inhabit poor land, at a diſtance from the rivers. There is a couple in Orangeburgh diſtrict, near the road that leads to Columbia from Orangeburgh, who lately had fifteen children alive out of ſixteen, and a fair proſpect of more. Another couple live in Darlington-county, fifteen miles from Lynch's-creek, who lately had thirteen children, and fifty-one grand children, all alive; and of their thirteen children, twelve were married at the ſame time.

The yellow fever raged in this city in the years 1700, 1732, 1739, 1745, 1748; but ſince the laſt mentioned year, nothing of the kind, of ſerious conſequence,* has taken place,

E   except

* Some perſons die almoſt every year, with the bilious fever, whoſe ſkin is yellow before or after death, and

except the malignant fever of 1792 and 1794;
which, though it refembled the yellow-fever
in many things, was entirely different in two
important particulars. It was not contagious,
nor did it affect any perfon who had, for any
confiderable time, been ufed to the air of
Charlefton.

Sundry perfons from the country were in-
fected with it in this city, who died on or
immediately after their return; but in no inftance was the difeafe propagated from them,
nor among the attendants on thofe who had
the difeafe in Charlefton. It was a fever
fui generis, but refembled the typhus icterodes of Sauvage. The whole mortality
from it, in 1792 and 1794, did not exceed
one hundred and fifty in each year.*

                                        Camp

and fome of whom difcharge black matter by vomiting;
but this is very different from what is commonly meant
by the Weft-India yellow-fever.

    * It is much to be regretted that regular bills of mor-
tality are not kept in Charlefton. To remedy this de-
fect, on a particular occafion, the fextons of the different churches were defired to give information of the number of perfons buried in their refpective burial grounds,
from which it appeared to the medical fociety, that between the firft of Auguft, 1792, and the 26th of October, of the fame year, one hundred and fixty-eight
white perfons were interred in the different burial grounds
in Charlefton. When it is confidered, that the typhus
icterodes began about the middle of July, and did not
difappear till the middle of October, of this fame year,
                                        1792,

September.	Intermittent fevers. Catarrhal fevers. Meafles. Angina ulcerofa. Croup.	Typhus icterodes. Catarrhal and rheumatic fevers. Dyfentery. Hooping cough.	Catarrhal fevers.
October.	Catarrhal fevers. Spafmodic colics. Intermittent fevers. Meafles.	Intermittent fevers. Croup. Small-pox.	Catarrhal fevers. Scarlatina. Intermittent feve
November.	Croup. Scarlatina anginofa.	Small-pox. Hooping-cough. Intermittent fevers.	Hooping-cough. Catarrhal fevers.

Camp fevers were, as ufual, attendant on the armies in the time of the late war. The fcarlatina anginofa was alfo common in Charlef-ton, in the year 1783, but attended with little mortality. The typhus icterodes of 1792 and 1794 was confined to ftrangers, and did not extend beyond the limits of this city. Thefe difeafes were, in a limited fenfe, epidemic; but, except the influenza, no ferious extenfive epidemic has taken place among us for the laft twenty years.

The annexed table, extracted from the journals of the medical fociety, will fhew, at one view, the general tenor of the difeafes that have occurred in Charlefton, for the five laft years.

It muft be highly agreeable to every bene-volent mind, that Charlefton is now more healthy than formerly, and likely to be more and more fo. With pleafure I anticipate, that in the courfe of the next century, our buildings will be extended into Afhley and Cooper rivers, as far as low water mark; that the adjacent marfhes will be banked in; the ftreets paved, and well provided with fewers; the bogs drained; the low grounds filled up; and the whole area of the city be firm, folid, high,

E 2 and

1792, and that Auguft and September are the moft fickly months of the whole twelve, the death of one hundred and fixty-eight perfons, in the courfe of eighty-feven days, in a city, whofe white population was about eight thoufand perfons, muft be deemed very moderate.

and dry land. Thofe who recollect the time
when ducks were fhot in a pond, which occu-
pied the ground on which the ftate-houfe is
erected—when a creek ran up to Church-ftreet,
and was croffed on a bridge, near where the
French church now ftands—when they ufed
to fwim over that fpot of ground which is now
Mr. Allfton's garden—when Water-ftreet,
which, at prefent, is high and dry, was almoft impaffable, will acquit me of being too
fanguine, when I indulge the hope, that our
grand-children will be lefs expofed to fevers
than we are.

It is a glorious exploit in a country, whofe
maladies chiefly arife from heat and moifture,
to redeem its metropolis from moifture, which,
of the two, is the moft plentiful fource of
difeafe. Whoever builds a houfe, fills a pond,
or drains a bog, deferves well of his country.*

It

* Our fellow-citizen, Captain Toomer, is entitled to
praife on this account; he has converted a very miry fpot
in Meeting-ftreet, into folid ground, and covered it
with houfes. Much remains to be done in this way, to
improve the health of Charlefton. The exiftence of a
pond in a city, is a reproach to its police. Efficient mea-
fures fhould be immediately adopted to drain or fill up
the low grounds. The ftreets fhould be paved, and the
fewers conftructed on a different plan. They ought to
be completely covered over, and extended on each fide·
to the neareft river: while fmaller ones, from every houfe,
fhould enter them near their top, and on a defcent. All
offenfive matter fhould be tranfmitted through thefe late-
ral fewers to the main one in the middle of the ftreet;
and

It is no small advantage to the inhabitants of Charleston, that they can, in the space of two hours, parry the heat of summer, by going to Sullivan's-island, where many invalids, especially children, have found a speedy restoration to health and strength. Our citizens have gained so much by frequenting this island, we may well wonder that is only three years since it began to be a place of summer resort.

Intermitting fevers are common to those who inhabit on or near to the banks of our rivers. On the other hand, by removing into the high and dry lands, three or four miles from the rivers, ponds, and mill-dams, fevers may, for the most part, be avoided. Of this a remarkable instance has lately occurred in St. Stephens, the inhabitants of which by quitting the swamps in summer, and fixing themselves in a new settlement, called by them Pine-Ville, have, for two years past, in a great measure, escaped the diseases which are common in the most sickly season of the year.

The swamps of South-Carolina terminate about one hundred and ten miles from the sea-coast; from thence westward the country becomes more hilly: the inhabitants are more ruddy, and in general more healthy.

The

and the whole so constructed, that as often as it rained, there would be a general purification of the city.

The tetanus is more common here than in
colder countries. Twenty-one cafes of it,
and moft of them fatal ones, have been reported to the medical fociety, between September, 1791, and Auguft, 1795: feven of
thefe took place in winter. Chronic complaints
are comparatively rare in this ftate. The
gravel, the ftone, the dropfy, the rheumatifm,
and the confumption occur much feldomer
with us* than with our northern brethren.
Fevers are our proper endemick: he who ef-
capes them has little elfe to fear. And much
may be fuccefsfully done for the avoidance of
them by prudent careful active perfons, who
ftudy their conftitutions, and obferve a gene-
rous medium between living too high and living too low.

Were it poffible exactly to contraft the
confumptions of New-England with the fevers
of South-Carolina, the inhabitants of both
would have nearly equal reafon to be fatisfied
with the place of their nativity. As to long
life our eaftern brethren have the advantage
of us. In proportion to numbers, as far as hif-
tory

---

* " In tropical countries, people are feldoml affected
with dangerous pulmonic difeafes; idiotifm and mania
are very uncommon: lunacy is almoft unknown: fcurvy
and gravel are difeafes feldom to be met with, and the
ftone fcarcely ever. I have known many Europeans fubject to the gravel at home who had no fymptoms of it
during their refidence in the Weft Indies."

Mofely on the difeafes of tropical climates, p. 112.

tory and obfervation warrant a comparifon,
there are as many of their inhabitants reach
85 as of ours who attain to 70.

Extreme old age, though not common, is
fometimes attained by our citizens, efpecially
by thofe who, in middle or early life, have
migrated from the cold northern countries of
Europe. A native of this city now refides
in it, at Amen corner, who is fuppofed by her-
felf and acquaintances, to be an hundred years
old. I have been well informed of feven or
eight others in different parts of the ftate, who
have reached, and in fome cafes exceeded that
period. A particular cenfus of the aged inhabitants of this city was taken by Captain
Jacob Milligan, in the year 1790, at the requeft of a worthy citizen, fince dead, from
which it appeared that there were then, in
Charlefton, 198 white perfons who were fixty
years of age, and one hundred of thefe were
upwards of 70, and one 108. Our white population, at that period, was about 8000.

This imperfect fketch of the foil, climate,
weather, and difeafes of South-Carolina, collected from our medical journal, my own obfervations for 22 years, and the information of
others, is refpectfully fubmitted to the fociety, with a requeft that each member would
freely point out wherein I am deficient, and
where

where I am miſtaken.   He who, in the ſpirit
of candor and philoſophy, corre&ts me in an
old error, or furniſhes me with a new truth,
deſerves, and ſhall receive my moſt grateful ac-
knowledgements.

David Ramſay.

CHARLESTON, S. C.
May 1, 1796.

www.ingramcontent.com/pod-product-compliance
Lightning Source LLC
Chambersburg PA
CBHW022032190326
41519CB00010B/1683